Giocondo Bocciarelli

Topics of Theory
of Neurosciences:
Behavior and Emotions

2

To Lucia and Elisa

ISBN 978-1-4092-0211-0

PREFACE

This work tackles topics that are thought enormous and impossible, but I have focused on the strong key problems and have done some answers.

This is essentially an essay of neurosciences, of theory of neurosciences. It is not of psychology, that is an applied science, according to me. It is not a popular manual of how to manage our emotions. Despite the aspect, it is not a philosophical paper, but it is obvious the great interest of these disciplines on it. But also of A.I., sociology, theology, and so on.

The models are described by a synthetic and unorthodox way. Some passages are very difficult to understand without a solid background (and an open mind too...). The first part is already very old (it was written in 1995) but it was left so to verify it in the time. The second part is of 2005.

G.B.

THE AUTHOR

Giocondo Bocciarelli is a doctor specialized in neurology and emergency at the University Hospital of Perugia – Italy.
He has a special interest in theory of neurosciences and emergency.

E.mail: *giocondo.bocciarelli@poste.it*

INDEX

6

Introduction

In the absurd event of an ancient Roman reappearing on the Earth today, and meeting then a car, he would probably look at it with wonder, and he would call it by a different way, perhaps "carrus sine equo". But the question of definitions, being at first nominalistic, becomes central as one tries to spot some great functions and their corresponding frames. One of the features of this "carrus", for instance, is that of moving and reducing its speed so much that it comes to a stop. But if our Roman does not master the concepts of deceleration due to lowering of power (decrease of fuel), and to internal (e.g. brakes) or external (e.g. slope) resistance, he would finally say, for instance, that it stopped only because the slope; whereas deceleration may have been due also to the other causes. Therefore his calculations upon part of the observed phenomena are not going to be right, and he will often draw some distorted general conclusions. Or else, if he thought it is only the accelerator that accelerates and decelerates the car, he would not determine the wheels' braking system, and he would consider it as linked to the same wheel without any particular distinct function. Nevertheless the wheel has several well distinguished structural subsets (suspension, axle shaft, brakes, tires and so on), and each of them has very definite functions (bearing, stabilizing, propellent, braking, steering and so on).

All this underlines to us how little mistakes in a general definition put seriously off the track particular hypotheses and a great number of those experiments based on that system. As an instance, we may here mention the thin and yet enormous difference in defining force between Aristotelian and Galilean mechanics: $f=mv$ changes in $f=ma$! We can imagine what is nowadays happening to us on such a complex subject like the "brain-mind" system.

8

1. A General Integrated Theory of Behavior

Even though scientists have carried out thousands of experiments over the years in all disciplines of behavior, and discovered a lot of laws and elaborated many theories, there is still a need for one clear general explanation of phenomena associated with behavior.

So far a precise and effective definition of the following processes has not been reached: conditioning, learning, behavior, intelligence, problem solving and knowledge. Researchers had been satisfied with vague definitions and thus a restricted area could be attacked in order to discover less complex laws and to elaborate limited theories. Each school has developed a particular experimental method and a precise model which have often been successful. But then they tended to enlarge their system towards all other psychological phenomena, and were generally unsuccessful.

Only recently the interdisciplinary approach has become dominant. But there are many difficulties as far as methods, exchange of knowledge and integration of theories concerned, because each researcher substantially remains attached to his original principles. Now all the generic, redundant, incomplete and erroneous parts of each theory must be eliminated and the three sources of behavior (genetic, environmental and mental) have to be integrated into a wider organic framework.

The early behaviorists tended to be influenced by an obsessive sense of objectivity; and when faced with the subjects of their experiments they tended to lose the overall view of the phenomenon, in favour of the narrow aim of solving the problem at hand, that is the relationships S-S and R-S. Modern scientists of conditioning, who are fundamentally *connectionists* according to Hill (1990, p. 23), have changed their theories very much and have been moving towards a more cognitive and genetic perspective (e.g., Rescorla, 1988; Staddon, 1983).

However, up to now they have made use of Skinner's reasoning thus dividing conditioning in two types, classical and operant (see Mackintosh, 1983, chap. 2). This is an important point in that it blocks the process of integration.

Moreover, it is impossible to understand the mechanism of a complex information processor such as the brain only by defining some of its inputs and intervening variables. In spite of Hull's (1943) ingenious attempt, behaviorists did not really want to study behavior and could not do it, but ended up employing it to try to understand learning.

On the other hand some important studies, e.g., Breland & Breland's (1961) and Seligman's theory of preparedness (1970), have pointed out the importance of the genetic approach carried forward by *ethologists*. Also neurobiologists, from Bard onwards (the "sham rage", 1928), have confirmed that some behavior patterns are "stamped" in the brain. But ethologists have not produced a clear and well-structured formal analysis of behavior.

Now it is evident that some phenomena cannot be explained without looking inside the mind. Today the *cognitive* approach is prevalent in psychology. It has taken on a new lease on life since the mind has come to be considered a biological processor of information. The result of this is that an S-R analysis of behavior is no longer considered very relevant.

However all the complex structures and procedures that are programmed on computers, and which are necessary for human reason in order to understand events, often can be performed more simply and rapidly by any brain.

Finally in this perspective we enter into the mind but not into the brain, since, according to many computational cognitivists, hardware structures are not important, but rather the way programs are constructed (Johnson-Laird, 1983, pp. 9-10). Nevertheless, it is not easy to agree with this point, because it assumes that all the information processors, both biological and artificial, are substantially equivalent. On the other hand a detailed and comparative analysis of the anatomy of the brains of various species and hardware of computers of various types demonstrates that there are many differences between them that are reflected in their behavior and working. So we risk building models that work only in the researchers' minds and computers. Therefore the structures of all processors must be studied and, at the same time, correlated to their behavior mechanisms. Other researchers, however, have been working in this direction (e.g., see Kosslyn & Koenig, 1992, and their concept of "wet mind", pp. 3-4).

1.1) Classification

```
RELATION LIFE=BEHAVIOR
     PREPARATORY PHASE
          HABIT
                GENETIC MEMORY
                EPIGENETIC MEMORY
          ADAPTATION=PROBLEM SOLVING
               NON-ORIGINAL SOLUTION
                    EXTERNAL
                         IMITATION
                         COMMUNICATION
                    INTERNAL
                         GENETIC MEMORY
                         EPIGENETIC MEMORY
               ORIGINAL SOLUTION
                    EXTERNAL=LEARNING
                         QUANTITATIVE
                              HABITUATION
                              SENSITIZATION
                         ASSOCIATIVE
                              CONDITIONING
                              IMPRINTING
                    INTERNAL=INTELLIGENCE
     CONSUMMATORY PHASE
VEGETATIVE LIFE
```

Fig. 1 - Behavior process classification scheme.

The RELATION LIFE or BEHAVIOR (Fig. 1) is the relationship between the individual and his environment over time. The general purpose of behavior is to ensure the survival of the individual, his growth and his reproduction, in any given environmental condition. Thus he has some vegetative needs that he tries to satisfy. The relation life is made up of a series of parallel sensations and actions which are out of phase and which form "paths" (see Sec. "The process model"). It is divided into two precise parts: the preparatory phase and the consummatory one.

The PREPARATORY PHASE is characterized by the need to obtain an object that is necessary for the "endosoma" (see Sec. "The structure model"), and that is satisfied by carrying out a pattern of behavior, e.g., fighting, courtship and predation, etc. In this phase there is a wide choice of paths, some of which are to be performed (rewarded) and others that should not be chosen (punished) in order to pass to the consummatory phase.

These behavior patterns can be modeled on the most recent habits, and thus derived from both *genetic memories* and *epigenetic* ones (see Sec. "The structure model"). So in this case the preparatory phase is in the HABIT state.

Nevertheless over time some major and sudden mutations can occur in the environment or in the individual too. This new unpredictable situation invalidates the most recent habit behavior and brings the individual-environment relationship to a crisis. Thus a *problem* occurs, i.e. how to restore the right relationship with regard to satisfying a precise need. So the individual will look for a solution that is different from his most recent habits. This solution might have been used in the past, but not recently. Therefore we can say that the preparatory phase is in a state of ADAPTATION (i.e. the present-day "problem solving" or "learning" in the wider sense), which is alternative to the habit state from which it is derived and towards which it tends. Its aim is to discover the path that leads to the consummatory phase, and when this path is stabilized, (i.e. often used in the same situations), it becomes a new habit. These solutions can be either original or non-original, in reference to the individual.

The *non-original solutions*, already existent, can be derived from outside the cerebrum, since it is connected to the environment, and this occurs either by IMITATION or by direct COMMUNICATION, or from inside by internal observation into *genetic* or *epigenetic memories*.

Otherwise, the individual can discover an *original solution* by himself, which may be totally new, at least for him. This also can come from two sources: inside or outside of the cerebrum.

The external solution occurs by way of trial and error in the real world. This is the mechanism of LEARNING, a process in which the cerebrum establishes some strong associations between certain sensations and actions in the real world (cf. Mackintosh, 1983, p. 77; Mazur, 1990, p. 2). Experience is not the correct term to use here if the concept of knowledge is still linked to it, because this would cause us to be guilty of circular reasoning as described by Bower & Hilgard (1981, p. 2). The associations can be non-adaptive, or without any changes in behavior, or not resulting in performance: e.g., omission training of Coleman (1975), the blocked performance by atropine in Finch (1938), "latent learning" of Tolman (1932).

Here as well, the mechanism can be divided into two procedural modes: quantitative in which response modification either decreases (HABITUATION) or increases (SENSITIZA-TION); and associative, where some new associations are formed among various elements, and is of two types, CONDI-TIONING or IMPRINTING.

The solutions elaborated inside the mind are the product of INTELLIGENCE. In this mechanism the cerebrum elaborates many mental trials and later some of these "attempts" will be checked in the real world. Up to now this process has been called "thinking" or "problem solving" or "cognitive learning", but the use of these terms demonstrates lack of clarity in the classical definitions. For instance, Kohler in his paradigm with apes (1927, pp. 46-48) did not actually realize that he was not dealing just with learning, nor with a mere cognitive process, but with a mixture of several adaptive mechanisms. In fact ape Sultan, in the first part of the experiment, carries out some trials in the environment (learning state) which, after proving unsuccessful, he discontinues. In spite of this he goes on thinking about possible solutions until he foresees the right one in his mind and then this solution is performed by him immediately in the real world ("insight" phenomenon seen from outside).

When the individual discovers and obtains the object of need, then the CONSUMMATORY PHASE is triggered. This phase is devoted to working on the object in the interface (see Sec. "The structure model") between environment and endosoma, e.g., eating, copulation, urinating (cf. Von Bertanlaffy, 1950).

At the end of the activity in the consummatory phase the new object of need, if it has entered the soma, is further worked on as far as the individual's VEGETATIVE LIFE is concerned. This occurs in parallel with the relation life and is completely automatic. It is made up of the main vegetative functions: digestion, reproduction, breathing, etc.; and in turn works on the objects up to the molecular level, which may be structural (e.g., proteins) or energetic (e.g., glucose). Moreover it has the function of internal micro-defence, e.g. with regard to bacteria.

Vice versa, if the object comes from the endosoma, and then is expelled during the consummatory phase, it is later free in the environment, concluding, in this way, the general cycle of the behavior process.

1.2) The process model

The process that puts into practice the above definitions has to be analyzed here (Fig. 2).

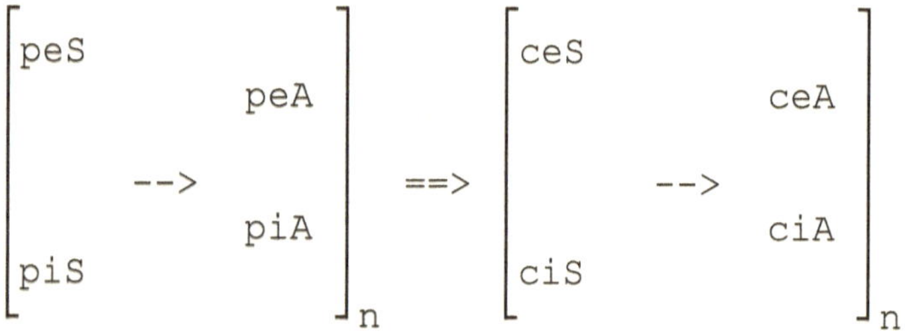

$$
\begin{bmatrix}
peS & & \\
& peA & \\
--> & & \\
& piA & \\
piS & &
\end{bmatrix}_n
\implies
\begin{bmatrix}
ceS & & \\
& ceA & \\
--> & & \\
& ciA & \\
ciS & &
\end{bmatrix}_n
$$

Fig. 2 - Process model scheme.

```
(p = preparatory; c = consummatory;
 e = external; i = internal;
 S = Sensation; A = Action;
 n = ordinal number)
```

A SENSATION (*S*) is the representation in a certain moment of a sensorial field of the environment or of the soma in the cerebrum's input units (see Sec. "The structure model"). The aim of every sensation is to prime a "good" subsequent action, and thus it essentially is a signal. Moreover every sensation causes either negative or positive feedback with regard to the previous action.

EXTERNAL SENSATIONS (*eS*) are sensations that refer to the environment. The individual has a series of eS that are made up of many elements that are difficult to know exactly. Hedonistically beautiful and ugly eS are defined by comparing them to others that are situated in the genetic and mass memories (cf. the "Innate releasing mechanism" used by ethologists, Lorenz, 1981, pp. 153-175). They are linked to a precise set of responses (piA and peA, see below) that marks each eS.

In the preparatory phase, an external Sensation (*peS*) is the survey of objects in the environment that can be either aversive or appetitive. The former stimulates an avoidance tendency and the latter, on the other hand, stimulate an approaching one. A peS can be a potential ceS (see below) or associated to another good or bad peS. During the preparatory phase the eS generally come from far ranging sensors: eyes, ears and nose.

In the consummatory phase, an external Sensation (*ceS*) is a gratifying sense of well-being regarding an object which is without doubt inside or outside of the soma. It can be either appetitive, i.e. the possession of the desired object (e.g., food) or aversive, i.e. the removal of an aversive one (e.g., shock, urine). In the consummatory phase eS usually comes from contact sensors: tongue, skin.

It should be pointed out that the object of a drive, which is sought by the "esosoma" (see Sec. "The structure model), might be different from the corresponding object of need that will be worked on by the endosoma. For example, a hungry baby will look for its mother's nipple, and the nipple is worked on during the consummatory phase, thus producing milk, which is the need of the endosoma and therefore of the vegetative life too.

An INTERNAL SENSATION (*iS*) is the representation of a soma's area, especially of the endosoma.

In the preparatory phase, the internal Sensation (*piS*) is a sense of inconvenience in the soma that can be either appetitive (e.g., hunger) or aversive (e.g., pain). It represents the individual's needs.

The individual associates certain piS (e.g., pain due to shock) with the corresponding peS (e.g., the shuttle-box) and he tends to react to the latter in order not to feel the corresponding piS.

During adaptation the parameters of the piS can change according to the situation. For example, a great increase in heart rate may occur after a shock because it is in support of a significant and immediate flight reaction. But if the adaptation has already occurred and if the solution does not imply doing heavy work, then the heart rate slows down in response to the same stimulus.

In the consummatory phase, the internal Sensations (*ciS*) are characterized by a gratifying sense from the endosoma that may be either appetitive (e.g., satiety, orgasm) or aversive (e.g., sense of well-being after pain). They terminate all the behavior patterns for that need.

An ACTION (*A*) is a set of commands that at a certain moment the cerebrum's output units (see Sec. "The structure model") give to the actuators which are, on the one hand, smooth muscles and glands, and, on the other, skeletal muscles. They can be contractions, relaxations and secretions, but all are integrated and coordinated. The goal is always to obtain a certain "interesting" sensation subsequently, and thus every action is of a selective and instrumental nature.

EXTERNAL ACTIONS (*eA*) are products of the skeletal muscles in response to the output units' commands which can come either from genetic memories or mass memories or may be partly new.

In the preparatory phase, external actions (*peA*) consist in setting a sensor and, if necessary, head and body, in one direction or manipulating an object. But each peA is both operant and perceptive at the same time. All peA are selective and their only aim is to reach the next interesting sensation (cf. Wickoff, 1952; Skinner, 1938). It is not important if this is reached by way of, for example, the hand muscles (e.g., adapting the fingers to a particular object's shape) or by the eye muscles. A peA can even be to remain motionless and to stare in a certain direction because the right sensation can only come from there. Moreover, even simply doing nothing can be very active and significant, e.g. a soldier who does not salute an officer!

In the consummatory phase, external actions (*ceA*) are those of the esosoma which work on the introduced object in the interface chambers in order to expel it from the soma, or to produce the substances directly usable by the endosoma (milk, sperm, bolus, etc.).

INTERNAL ACTIONS (*iA*) are actions towards particular areas of the soma. They go along with the external ones, and often make the cerebrum aware of the endosoma's state. For instance, hypoglycaemia becomes a *piA* (internal action in the preparatory phase) of hunger by way of stomach contractions. Moreover piA enable the soma to carry out the peA, e.g. all vegetative changes caused by fear which are in support of flight from unpleasant sensations.

In the consummatory phase, internal actions (*ciA*) follow rewards (*ceS*) but then, during adaptation, they can be anticipated in order to allow the beginning and the maintenance of the associated ceA: e.g. erection > copulation, salivation > chewing (cf. the atropinized dog that refuses food, Finch, 1938).

Actions are an integrated, complex and teleonomic set of responses of both the endosoma and the esosoma, and not simply tachycardia or salivation or lever pressing, as many conditioning theories have interpreted them. For example, flight is the peA of fear whose piA consists of tachycardia, shaking, etc. In Mowrer's "two-factor theory" (1947) the stimulus elicits the state of fear which in turn activates the skeletal response. Actually peA and piA occur at the same time; they are two aspects of one single response. However we can feel fear after carrying out the peA because some responses are slower to come about and to be noticed by the endosoma than others. For example, a man is driving calmly; suddenly he meets an obstacle on the road and applies the brakes firmly (peA), but only afterwards does he feel his heart beating very fast (piA > piS).

An important point is that in the case of indecisiveness two (or more) pA can be activated, bringing about a state of internal conflict (cf. Lorenz, 1981, pp. 243-245), in which one pA may try to produce an effect and the other pA an effect completely opposite at the same time.

The PREPARATORY PHASE is characterized by the discovery and the obtaining of the object to be worked on or by creating the conditions for the priming of the consumma-tory phase. The preparatory phase: (1) can be either appetitive or aversive; (2) allows the adaptation to take place even though it may be formed by some genetic chunks; and in addition, (3) it is very highly developed in humans as well as being complex and indirect with respect to the consummatory phase.

The law of adaptation involves always starting with a punishment, which can be either "appetitive omissive" (e.g., lack of food) or "aversive effective" (e.g., shock), and termi-nating with the last reward. Reward can be either "appetitive effective" (e.g., possessing food) or "aversive omissive" (e.g., absence of shock). This definition is clearer than Skinner's (1953, p. 73) because the terms "reinforcement" and "rein-forcer", along with their positive and negative modifiers, are sometimes confusing since they have been used with various meanings.

Some examples of the preparatory phase are the behav-ior patterns of hunting and courtship, which, in the consumma-tory phase, are followed by eating and copulation respectively. Pecking and biting are not the initial action of the consumma-tory phase but rather the last one of the preparatory one.

In some cases the preparatory phase may not be fol-lowed by the corresponding consummatory one. For example, in the behavior patterns concerning play and exploration the individual has not got a vegetative need but, nevertheless, searches in the same way without activating the ceA. Another example of this phenomenon is when an individual is hunting, and this is not followed by eating or, in the case of courtship, mounting which is not followed by copulation (cf. Lorenz, 1981, pp. 325-335). This is a good example how different mechanisms (e.g., genetics, imitation and learning) can converge to obtain the correct behavior. For example, in rhesus monkeys the mounting by the male over the female's posterior legs, which is the last part of the preparatory phase, must be acquired during its early social life by observation and play. In fact, males reared in isolation are neither capable of doing this adequately nor of copulation (Harlow, 1962).

The CONSUMMATORY PHASE is the interface period between environment and endosoma. It is distinguishable in that it: (1) consists of actions which precede those of the endosoma and are coordinated with them; (2) occurs in the interface chambers; (3) has paths which are standard and genetic; (4) can be inactive (e.g., rest or sleep), or active introductive (e.g., food, water, sperm), or active expulsive (e.g., feces, urine, fetus); (5) is the second and final phase of relation life, that terminates a particular behavior, thus satisfying a certain need.

In humans the consummatory phase assumes some complex characteristics derived from civilization (e.g., feasts), and although it is more separated from important preparatory behavior patterns that really are of support to it, it is always present in everyday life. For example, civilized man finds it is easy to get food, the difficult thing is to make money in order to acquire it!

This phase as well can run independently without a need or other drives: e.g., a small monkey sucking a breast without milk, the "rage copulation", etc.

The above definitions regarding the preparatory and consummatory phases are different from other ones, for instance the nebulous one of Konorski (1967), the one more precise of Woodworth (1958, p. 223) and the most important one of ethologists (Craig, 1918). This last definition states that motivation leads the individual to be in a state of continuous agitation, that is "appetitive" or "searching" or "preparatory" behavior. For example, a hungry eagle apparently flies aimlessly over a valley until it meets a "key stimulus" (e.g. a rabbit) which triggers the consummatory act, i.e. attacking, capturing and eating. Whereas, according to the model presented here, the consummatory phase begins only when the food is in its mouth without the possibility of being threatened and thus lost (ceS). From this perspective all the stages concerning the sighting of potential reward and its actual capture can be either in a state of adaptation or in one of habit: e.g. a lion may have learned that zebras are always near a river...

The BEHAVIORAL PATH is a series of cycles of sensation and action. In the adaptation state the right path must be discovered that leads to the ceS, thus solving the problem. This occurs by producing and selecting various S-A cycles that seem to be correct. Some of them can then be eliminated because they are recognized as being superfluous, even if they were useful at the beginning in order to reach the goal. Chunks of some paths can be copied, divided, modified, erased or even linked to form a new path.

During the adaptation state the individual records in his mass memory a lot of paths that are not all right or useful for that action. But in the future he may refer back to them using these associations for different situations. This phenomenon is called "latent learning" (Tolman, 1932), but this is not the right term for it. In fact, the association is latent but the mechanism that brought it about is always evident.

The paths of the preparatory phase are a mixture derived from different sources: conditioning, intelligence, genetic memory and epigenetic memory...

Sometimes it is impossible for an individual to solve the problem or it is difficult to discriminate the most correct path from those available, and so some paradoxical, strange or negative phenomena may occur: autoshaping, partial reinforcement effect, omission training, learned helplessness and superstition, etc.

Moreover, many potential paths inscribe themselves on a behavioral network of possible events which can be analyzed by using a cognitive approach.

1.3) The structure model

Now we must describe the major elements presented in the following diagram regarding the individual, the brain and the mind, as a support of the general mechanisms expressed above (Fig. 3).

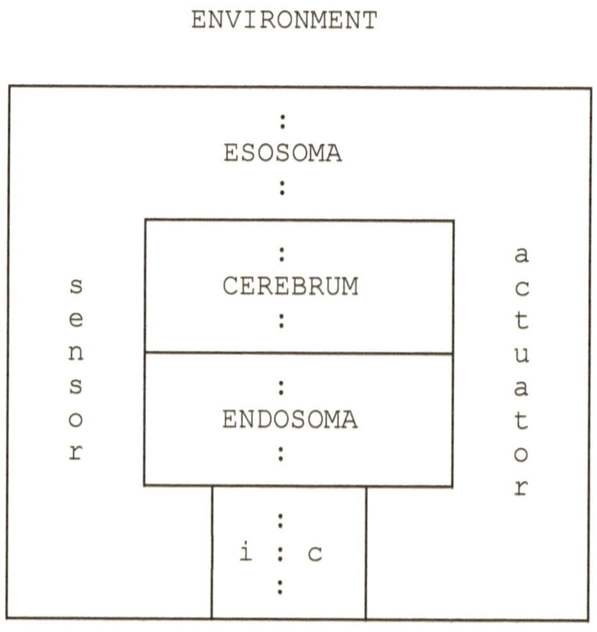

ENVIRONMENT

Fig. 3 - Structure model diagram.

(ic = interface chamber)

Every highly structured biological being, i.e. humans and superior species of animals, can be divided into two fundamental parts: the CEREBRUM, i.e. the brain plus spinal cord normally known as the central nervous system, and the SOMA, i.e. the rest of the body. The soma in turn is divided into two parts: the ESOSOMA, i.e. muscles and bones, that regards the relation life; and the ENDOSOMA, i.e. viscera, that concerns the vegetative life. They work in parallel and are complementary and coordinated each other. Both the esosoma and the endosoma have an input section made up of SENSORS (molecules, cells, organs such as eyes) and an output one, consisting of ACTUATORS (muscle fibers, muscles, glands). The soma is the sensory-mechanical apparatus of the individual and the cerebrum is his information processor, which checks and commands all of the soma's elements. The cerebrum is in contact with the environment by way of the esosoma.

The INTERFACE CHAMBERS, which are a bridge between the endosoma and the environment, are the channels of the soma at the boundary of the eso- and endosoma: mouth, vagina, nipple-ducts, rectum, etc. Normally they are closed but can extend themselves even very much in the presence of an object that must be worked on during the consummatory phase (e.g. penis, fetus, nipple, bolus, urine, feces). They often have smooth and skeletal muscles whose actions are coordinated in order to work on and introduce or expel objects.

The cerebrum has a structure that is made up of: (1) external input units, (2) internal input units, (3) external output units, (4) internal output units, (5) processing units; (6) genetic sensor memory units, (7) genetic actuator memory units, (8) epigenetic sensor memory units, and (9) epigenetic actuator memory units.

The cerebrum elaborates (processes) two types of information: sensations and actions, and so it has a sensorial section and an actuative one. The information patterns are elaborated independently and in parallel. This does not mean that in order to have a correct analysis of behavior the out of phase scheme (Fig. 4) must not be used.

```
... eS1 --> eS2 --> eS3 --> eS4  ...

 ... eA1 --> eA2 --> eA3 --> eA4  ...

 ... iA1 --> iA2 --> iA3 --> iA4  ...

... iS1 --> iS2 --> iS3 --> iS4  ...
```

Fig. 4. Cerebrum activity scheme.

```
(e = external; i = internal;
 S =  Sensation; A = Action)
```

During sensation each input unit, already connected to the soma's sensor and thus to the environment or endosoma, is in turn connected to the memory. During action the same thing happens with other memories and output units, that in turn are connected to their corresponding actuators of the esosoma which act in the environment, or to those of the endosoma.

The recording and the maintenance of all sensations and actions and their paths, the wrong ones as well, take place in various memories. The genetic memories of behavior are ROM-like (read only memory) and are common to all individuals of that species and change only by genetic mutation.

The epigenetic memories are mass memory-like and record the individual's experience and change continuously.

During learning the memory units are in contact with the environment and the soma through the input and the output units. Whereas during the mechanism of intelligence a part of the cerebrum is not in contact with the outside (i.e. the soma and the environment). Indeed the processing units exchange information between the memory units and the input units (or a hypothetical working memory). In this way the cerebrum can simulate reality and all its possible manipulations without moving matter and without energy consumption (cf. Kosslyn, 1983). Then the hypothetical correct action or path will be transferred to and verified in the real world by the esosoma.

28

2. Emotions-Feelings: Again, what are they? And how many are there?

Although it is now clear that most of the exact aspects of emotions have already been identified in one theory or another, the questions posed in the title still apply today. One has the impression of going round the problem in circles but still without grasping the structure of the process and, therefore, without having an overall picture. The great theories of the past tackled the issue from physiological expressions (e.g. Schachter and Singer, 1962), from facial expressions (e.g. Ekman, 1982), with linguistic analyses (e.g. Johnson-Laird & Oatley, 1989) and then with meticulous analysis of appraisal after cognitivism became domineering (e.g. Schorr, 2001). Lastly, and rightly so, we are now turning increasingly towards research in the neurosciences in order to identify the structures of the brain that produce these phenomena (e.g. Panksepp, 1998; Davidson, Scherer & Goldsmith, 2003). The need remains, however, for a correct guiding criterion in order to understand precisely what emotions are, so as to know exactly which are and which are not emotions, and to define the middle level between the upper phenomenological-behavioral one and the lower neurobiological one in approaching these studies (cf. Frijda, 2004, p. 171). The aim of this paper is precisely to focus on the problems preventing the solution from being reached and to provide a consequential, original and complete solution.

These problems are:

1) Emotion is a behavior and without a valid behavioral theory behind it, it is impossible to have a valid theory on emotion. This work is based on a definite theory of behavior that I exposed in the previous chapter.

2) There has always been confusion among needs, motivations, feelings and emotions, "emotional response", "bodily response", etc. Such words as love, disgust, good

/bad, positive and negative, and many others are used in a very ambiguous way not only in everyday life but also in the scientific literature. As is the case in natural sciences (for example, the concept of *force*), it is now unavoidable to clarify these definitions precisely and synthetically.

3) Lastly, although emotions are "pre-packaged" actions, they try to respond to a logic of adaptation on the part of the individual, and can, therefore, be analysed functionally on the basis of three precise axes of interpretation, that we shall examine later.

2.1) What is an emotion?

First of all, we must mention the still present century-old controversy between James and Cannon (James, 1884; Cannon, 1929), in which these authors unwittingly posed a fundamental question that is not among those normally considered, that is to say: are emotions basically sensations or actions? (cf. Izard, 1990; Scherer, 2000, pp. 154-156)

Before James, the first traditional and common sense theories considered emotions above all as sensations, followed by a response. As we can interpret and express, these sensations were visceral states and the responses to them were voluntary acts: I see a bear > "I am scared" > I run away (see Fig. 5).

Afterwards, James continuing along the same basic line, became fully conscious that emotions in this outlook are sensations, but: "emotions are embodied, that they emanate from the physical self and not from some 'spiritual' source outside a person's experience of self" as in the folk version (Barbalet, 2004, p. 221). Then he observed that these sensations must be the result of an "act", and very simply – but causing quite a clamour - reversed the process (combining, among other things, autonomic responses with skeletal responses): I see the bear > I tremble / I run away (note that these two types of responses were not together in paradigms he did) > I am scared.

In spite of this, his paradox was exactly here: the voluntary response came before the emotional sensation and, therefore, in order to escape criticism he willingly minimized it in some passages (cf. Ellsworth, 1994, p. 225, about the James' replay to Dr. Worchester's objection). For this reason, in recent decades the new cognitivists have substantially split up the response into an autonomic response, followed by the sensation-emotion, that is the feeling, and a voluntary response, preceded by a second appraisal (cf. Lazarus, 1991). In this way, moreover, the second and unappreciated aspect of these theories is highlighted, that is to say, either the total separation of the autonomic responses and the skeletal ones (cognitivists), or their merger without maintaining their diversity (James).

Cannon, on the other hand, was the unconscious but firm upholder of a radically different concept, that emotions are essentially actions, although no great distinction was made between external voluntary actions and autonomic ones.

THEORIES	PHASES					
Folks	Perception			"Emotion"		Skeletal R
James	Perception		Autonomic R or Skeletal R	"Emotion"		
Cognitivists	Perception	Appraisal 1 (Emotion)	Autonomic R (Emotion)	Feeling (Emotion)	Appraisal 2 (Emotion)	Skeletal R (Emotion)
Cannon	Perception			Autonomic R + Skeletal R		

Fig. 5 - Scheme of emotions' phases.   (R = response)

Thus, according this line but in a new framework, an EMOTION is a simple, archaic and genetic set of integrated and purposeful movements and secretions, that is inscribed in the ROM MEMORIES. That is they are pre-made actions! That is they are compound of *internal Action* and *external Action* (see Sec. "The process model"). Running away <u>is</u> the fear! Notwithstanding, these actions often cannot be implemented in full.

They are SUPERIOR INSTINCTS, that is, behaviors, stereotyped motor programmes from the Preparatory Phase, at least in superior mammals, while we consider the functions of the Consummatory Phase (breathing, eating, etc.) INFERIOR INSTINCTS.

A FEELING is nothing more than a lighter emotion, that is not transformed immediately into direct, strong action but does have an important influence on reasoning and facial expression. Thus, the sensory aspect is secondary, not as the word would have us believe! It is the other name for emotions, an important but misleading aspect that struck researchers working along James' line. More than anything else it is a *motion*, "a tendency to ...". The more an individual is calm and reflective, the more easily he will adopt complex and structured actions. These actions can also be INTELLIGENT, even if there is always a certain underlying predominant feeling towards a given situation, that limits and conditions the unconscious production of certain solutions rather than others, in the logical direction imposed by feeling.

Like all targeted actions, emotion obviously has an aspect towards the body (the *internal Action)* and one towards the environment (the *external Action*) which are integrated, parallel and synergic one to the other. That is to say, one automatic and within the body and one voluntary and external to it, that induces the body to move and to manipulate the environment. It is clear, therefore, that James was also right in a way: if <u>all</u> the expressions of an emotion, e.g. of fear, are eliminated, that emotion disappears. But we must be careful:

fear means running away, but often we do not want to do it immediately, and while the internal Action is activated directly (heartbeat, tightness of the stomach, etc., but also the use of the skeletal body leading to increased muscle tone and specific postures and facial expressions), the external Action is presented to the Voluntary Decision-making System, which decides whether to implement it or not, or which chooses another emotion or a motion produced cognitively by thought. Indeed, at this level there are *motions*, that is to say, proposals of external Actions of cognitive or emotional origin. For this reason, there may also be contemporary and opposite activations in the soma, and therefore the occurrence of a state of internal conflict that generates stress. It is like driving a car, accelerating while the brake is on: the car does not move, but the effort is imposing. If, on the other hand, the feeling becomes very strong (that is an emotion arises in the classic sense with an escalation), this motion will have such force that takes over the individual completely, and will be put into practice immediately and with decision, including its external Action part.

Over the millennia, evolution has selected these actions in the life of mammals and of man in particular, for the purpose of objectifying need, adapting it to a particular situation. If I do not choose something, how can I sooner or later satisfy my need? The brain detects the molecular and energy or defensive deficits, that is the *needs* of the body, by means of certain more or less clear and specific *internal Sensations*, that, therefore, become *motivations*. But then, through the esosoma, it must choose, find and conquer the specific objects that can potentially satisfy these needs. Thus, in the case of hunger, this means having a specific food. Lions do not eat grass but meat, in the form of a zebra, for example. But which zebra, and where and when? Without this automatic process – except for today's scientifically super-evolved human species – how does one know what to eat, where it is and how to obtain it?

The after-Sensations (internal Sensations of feelings) that record automatic activations following an internal Action are epiphenomena of emotions (e.g. the internal Sensation of tachycardia) that, however, also serve to make the individual aware of his emotional state, and to press him to act in a way that will enable him to reduce tension.

The Sensory Emotional Circuits evaluate the environmental Stimulus, the *external Sensations*, comparing them with genetic *external Sensation patterns* stored in the ROM MEMORIES or with earlier similar external Sensations and thoughts also stored in MASS MEMORIES. If they are found to correspond, the Actuative Emotional Circuits that produce the emotion action are activated automatically, unconsciously and directly. All these processes take place in parallel and are, of course, constantly up-dated and influence one another.

Within this framework, therefore, it is logical to interpret MOOD as the prevailing feeling over a certain period of time and PASSION as a very strong feeling with respect to an object or a very precise aim, that may even last for a long time (many years) and is perhaps reinforced, from time to time, by various suitable emotions.

2.2) How many emotions are there?

At this point we can say that if emotions are well targeted and integrated movements, they are of a well-defined number and have definite functions (for the debate on basic emotions see e.g. Sabini & Silver, 2005; Solomon, 2002). For this reason, there must be some logical axes for interpreting them that define them rationally. In my opinion, these can be clearly identified by means of functional analysis. They are: "beautiful" /"ugly", superior /inferior, "pre-" /"post-", that is to say: aesthetics, power ratio, verification.

The **aesthetic** point of view is the first feature of an object or a situation. It is the recognition that a given thing is significant for a certain need, on the basis of specific characteristics. Obviously, there is the usual directionality: keeping something beautiful /escaping something ugly, getting rid of something ugly /looking for something beautiful. The distinction between having something ugly (e.g., pain) and not having something beautiful (e.g., not obtaining food) is purely formal and linguistic. A feeling (e.g. an ugly feeling) remains, that differs only in terms of intensity and urgency, but in the long term it is always very negative (pain is more urgent but also unsatisfied hunger will lead to death!). We should not confuse internal Sensations (pain /hunger) with the external Sensations with which they are associated (e.g. fire /food), whether present or absent.

Then the perceived **power ratio** between the individual and the "other" (person, animal, object, circumstances, groups, god, etc.) is essential for determining whether I can have something that, although it is beautiful, might not be achievable (as we all know very well). So what is the point of making an effort? The more this ratio is unbalanced the clearer is the emotion, the more this ratio is balanced the more two conflicting emotions are rising.

Finally, a critical event is an effort, a struggle that can decide the win or the loss in a certain affair, often without reaching the extreme consequences (e.g., the threat that forces somebody to flee). The main function of a **POST-critical event** emotion is to sanction a possible change in a PRE-emotion associated with a specific Stimulus. Indeed, the post-emotions occur after a significant event, and do not contribute towards its occurrence. So what is their purpose? Do they function merely as a release? (cf. Frijda, 1994). For example, if, having come across a specific object such as something that has attacked me and caused me to run away many times, this time I fight and win (critical event), then this will make me happy and the emotion associated with it in the future will no longer be fear but aggressiveness. These emotions are, indeed, often characterised by surprise, and the more the event becomes customary the less strong the emotions will be. Further proof is provided by the fact that in all PRE-feelings one looks towards and thinks of the future, while with POST-feelings one looks back at and thinks of the past, of the fact that has occurred. In addition, obviously, they mark the object in question positively or negatively in an aesthetic sense.

PRE-emotions have also been called DRIVES, MOTIVATIONS, EXPECTATIONS, while POST-emotions have been called GRATIFICATION and FRUSTRATION. It should be stressed, on the other hand, that PUNISHMENT /NEGATIVE REINFORCEMENT, REWARD /POSITIVE REINFORCEMENT are external Sensations, not actions.

		PRE-	POST-	
Ugly	inferior	**Fear**	**Disappointment**	ugly-inferior
	superior	**Aggressiveness**		
Beautiful	inferior	**Desire**	**Joy**	beautiful-superior
	superior	**Love**		

Fig. 6 – Synoptic table of emotions and their axes of interpretation.

There are four PRE-emotions: fear, aggressiveness, desire and love.

In FEAR (that is to say ANXIETY, ANGUISH, PANIC, TERROR, clearly in order of intensity and escalation), something ugly has or is about to come along: I am inferior and I can avoid it. Fear is escaping, flight in an open field (= external Action, the internal Action is evident: tachycardia, trembling, etc.). A far-off external Sensation that recalls a frustrating one will be quickly escaped from. In ANXIETY the negative external Sensation may not be very obvious, but on analysing it carefully it can always be found, except in the case of strictly pathological anxiety that is disconnected from appropriate and proportional Stimuli.

AGGRESSIVENESS (that is to say IRRITATION, ANTIPATHY, CONTEMPT, HATE, ANGER, RAGE, FEROCITY, FURY): something ugly has or is about to come along but I am superior, so I threaten, attack, fight and destroy it. The same applies to somebody or something preventing me from obtaining the beautiful thing I want, for example food (cf. Berkowitz & Harmon-Jones, 2004). In social contexts there is also a communicative aspect: he should not have taken the liberty of doing that within the group in which I dominate!

LOVE is the feeling of giving something beautiful or of getting rid of something ugly. It is probably substantiated in the Parental Instinct Circuit. All the adjectives used unconsciously in situations of that kind recall that relationship. I am superior and will be happy when I have given something beautiful. It is often the fruit of joy. The corresponding emotions are smiling, caressing, hugging, giving. It is understood here as a basic emotion. Romantic love is a complex emotion (that is a mixture of love, desire and joy, with often some ugly ones) the object of which is sexuality, not to mention Christian love that also has other fundamental connotations such as ethical and metaphysical ones (cf. Djikic & Oatley, 2004).

DESIRE (SEARCH, HOPE, INTEREST), on the other hand, is wanting to receive beautiful things that I do not have since I am inferior. It means approaching with anticipation, wagging one's tail and imploring. It is seeking, working to obtain something (cf. Silvia, 2005).

There are only two POST-emotions: joy and disappointment.

JOY or HAPPINESS is when an individual has won and a PRE- feeling has been satisfied, no matter whether it is by having or giving or destroying or escaping! It up-dates the positive mark of that external Sensation as a possible and good object of consummation. This emotion is expressed in the form of skipping, laughter, specific mimicry, dancing, feasting, etc.

At this point, we have to make a note about "positive" emotions. They were ignored in all the old models of theories, because they are very evasive and nebulous without a solid framework of interpretation when analysed. In fact, only recently have they been taken in consideration, not regarding their meaning, but rather for their capability of adaptation, that is, for their appropriateness and moral judgement (see e.g. Fredrickson, 2004; Solomon & Stone, 2002; Ben-Ze'ev, 2000).

In DISAPPOINTMENT (or MORAL SUFFERING, SORROW, SADNESS, MELANCHOLY, DEPRESSION, DESPERATION), something ugly has come or is coming, but one is too inferior and is unable to escape it: *les jeux sont faits...* In its physiological form it serves the purpose of divesting our desire from an object that is no longer beautiful and /or that cannot be obtained, enabling us to pass to other ways of meeting a need. When it is very strong, it concerns life itself and serious depression sets in, leading to abandoning oneself to death (in animals) or to even actively seeking it (in man). Man blames himself morally for the fact. He keeps thinking of the past, of the ugly fact: post-emotions. He moves away from the object, and for some time solitude is sought. He is weak, submissive and withdrawn; he cries, etc.

It is clear that physical PAIN is a Sensation, a different one from those that occur in fear and in depression (or moral pain, although they are often associated), and this has also been demonstrated by experimental data as far back as the time of lobotomy, which eliminated the emotional aspect while maintaining the sensory aspect which was no longer disturbing (cf. Damasio, 1999, chap. 2). Furthermore, pain can be easily associated with anger! It is an internal Sensation not only of the visceral body but also of the sensors that are found throughout the body, except for the brain, of course.

SURPRISE, STARTLE and WONDER are not emotions but a sudden change from one emotion to another, in reference to Sensations the second of which (in order of time) is unexpected in relation to the first (for example, desire > fear: I open a box thinking there is a beautiful clothes whereas I find a snake!), and so this change may be either beautiful or ugly. Similarly, DISGUST is not an emotion but, as in the original Latin meaning, a non-emotional action expressed mainly facially, a lower instinct opposite to that of eating (and then swallowing), as a reaction to an ugly external Sensation of taste (and then vomiting). When it is understood as being MORAL REPUGNANCE or CONTEMPT, it is a sort of aggressiveness (cf. Royzman & Sabini, 2001).

FRUSTRATION, ENVY and JEALOUSY are often a complex situation, made up of more than one feeling of inconvenience (depression, anger, fear) that alternate and stimulate each other. It is also possible to detect emotions in man due to cognitive reasons that we could call meta-emotions, e.g. IRONY, when something absurd occurs or is hypothesized (e.g. I laugh seeing a mouse threatening to beat up a lion in a cartoon), or social emotions of fear, such as SHYNESS, or of aggression, such as DERISION.

Doubt as to whether something is negative is not due to fear but to a CONFLICT between two or more feelings. It is possible to have more than one feeling towards an object at the same time, and they may be conflicting! What is more, an emotion easily spreads to objects that have nothing to do with it or even to what could be useful. The action is discharged onto someone else, like a person who ruins his home rather than facing his enemy, or takes it out on his wife... since these are inferior. Or an emotion referred to a more important *behavioral path* becomes another emotion in an inferior *cycle* (see Sec. "The process model"). For example, if I am scared of arriving late at an important meeting with my boss, I get angry with the little old woman driving the car in front of me who will not let me overtake her! Lastly, PATHOLOGICAL feelings are the result of dissonance between the objective problem and the way in which the emotional person perceives it, which is often completely distorted. For others, the Emotional Circuits may also be hypertonic and /or completely illogically disconnected from the corresponding Stimuli (e.g. a person who laughs at a funeral).

Conclusions

In this little book I have drawn a general framework of the behavior process and than of emotions. This framework according to me is sufficiently clear but I am ready to do every explanation, in depth analysis and replay to any criticisms.

However there is a crucial topic that we have absolutely to clarify soon: the concept of *love*. It is a feeling so complex, widespread and fading that has to be examined separate. Other emotions, specially the ugly ones, are more easily to face.

Love involves ideas, believes, behaviors, studies that go from biology and neurosciences (pharmacology, anatomy, neurology, etc.) to psychology until philosophy, arts and theology! But we have to simplify, and my approach will be to enucleate the basis of it regarding the theory of neurosciences.

References

BARD, P. (1928). A diencephalic mechanism for the expression of rage with special reference to the sympathetic nervous system. *American Journal of Physiology, 84,* 490-515.

BARBALET, J. (2004). Hypothesis, Faith, and Commitment: William James' Critique of Science. *Journal for the Theory of Social Behaviour, 34* (3), 213-230.

BEN-ZE'EV, A. (2000). "I Only Have Eyes For You": The Partiality of Positive Emotions. *Journal for the Theory of Social Behaviour, 30* (3), 341-351.

BERCKOWITZ, L. & HARMON-JONES, E. (2004). Toward an Understanding of the Determinants of Anger. *Emotion, 4* (2), 107-130.

BOWER, G. H. & HILGARD, E. R. (1981). *Theories of learning.* Englewood Cliffs, NJ: Prentice-Hall.

BRELAND, K. & BRELAND, M. (1961). The misbehaviour of organisms. *American Psychologist, 16,* 681-684.

CANNON, W. B. (1929). *Bodily Changes in Panic, Hunger, Fear and Rage.* New York: Appleton-Century.

COLEMAN, S. R. (1975). Consequences of response-contingent change in unconditioned stimulus intensity upon the rabbit (*Oryctolagus cuniculus*) nictitating membrane response. *Journal of Comparative and Physiological Psychology, 88,* 591-595.

CRAIG, W. (1918). Appetites and Aversions as Constituents of Instincts. *Biological Bulletin, 34,* 91-107.

DAMASIO, A. R. (1999). *The feeling of what happens: Body and emotion in the making of consciousness.* New York: Harcourt Brace.

DAVIDSON, R. J., SCHERER, K. R. & GOLDSMITH, H. H. (Eds.) (2003). *Handbook of affective sciences.* Oxford University Press, New York.

DJIKIC, M. & OATLEY, K. (2004). Love and Personal Relationships: Navigating on the Border Between the Ideal and the Real. *Journal for the Theory of Social Behaviour, 34* (2), 199-209.

EKMAN, P. (1982). *Emotion in the Human Face.* Cambridge: Cambridge University Press.

ELLSWORTH, P. C. (1994). William James and Emotion: Is a Century of Fame Worth a Century of Misunderstanding? *Psychological Review, 2,* 222-229.

FINCH, G. (1938). Salivary conditioning in atropinized dogs. *American Journal Physiology, 124,* 136-141.

FREDRICKSON, B. L. (2004). The broaden-and-build theory of positive emotions. *Philosophical Transactions of the Royal Society of London / Series B – Biological Science, 359,* 1367-1377.

FRIJDA, N. H. (1994). Emotions are functional, Most of the time. In Eckman, P. & Davidson, R. J. (Eds.), *The nature of emotion: fundamental questions* (pp. 112-122). New York: Oxford University Press.

FRIJDA, N. H. (2004). Emotions and Action. In Manstead, A. S. R., Frijda N. H. & Fischer A. (Eds.), *Feeling and Emotions – The Amsterdam Symposium* (p. 171). Cambridge: Cambridge University Press.

HARLOW, H. F. (1962). The Heterosexual Affectional System in Monkeys. *American Psychologist, 17,* 1-9.

HILL, W. (1990). *Learning.* New York: Harper Row.

HULL, C. L. (1943). *Principles of behaviour.* New York: Appleton-Century-Crofts.

IZARD, C. E. (1990). The Substrates and Functions of Emotions and Feelings: William James and Current Emotion Theory. *Personality and Social Psychology Bulletin, 16* (4), 626-635.

JAMES, W. (1884). What is an emotion? *Mind, 9,* 188-205.

JOHNSON-LAIRD, P. N. (1983). *Mental models.* Cambridge, UK: Cambridge University Press.

JOHNSON-LAIRD, P.N. & OATLEY, K. (1989). The Language of Emotions: An Analysis of a Semantic Field. *Cognition and Emotion, 3* (2), 81-123.

KOHLER, W. (1927). *The Mentality of Apes.* New York: Kegan-Trench-Trubner.

KONORSKI, J. (1967). *Integrative activity of the brain: An interdisciplinary approach.* Chicago: University of Chicago Press.

KOSSLYN, S. M. (1983). *Ghosts in the mind's machine: Creating and using images in the brain.* New York: Norton.

KOSSLYN, S. M. & KOENING, O. (1992). *Wet Mind: The New Cognitive Neuroscience.* New York: Macmillian.

LAZARUS, R. S. (1991). Progress on a cognitive-motivational-relational theory of emotion. *American Psychologist, 46,* 819-834.

LORENZ, K. (1981). *The Foundations of Ethology.* New York: Springer-Verlag.

MACKINTOSH, N. J. (1983). *Conditioning and associative learning.* Oxford: Oxford University Press.

MAZUR, J. E. (1990). *Learning and behaviour.* Englewood Cliffs, NJ: Prentice-Hall.

MOWRE, O. H. (1947). On the dual nature of learning – A reinterpretation of "conditioning" and "problem-solving". *Harvard Educational Review, 17,* 102-148.

PANKSEPP, J. (1998). *Affective neuroscience.* New York: Oxford University Press.

RESCORLA, R. A. (1988). Pavlovian conditioning - It's not what you think it is. *American Psychologist, 43,* 151-160.

ROYZMAN, E. B. & SABINI, J. (2001). Something it Takes to be an Emotion: The Interesting Case of Disgust. *Journal for the Theory of Social Behaviour, 31* (1), 29-59.

SABINI, J. & SILVER, M. (2005). Ekman's basic emotions: Why not love and jealousy? *Cognition and emotion, 19* (5), 693-712.

SCHACHTER, S. & SINGER, J. E. (1962). Cognitive, social and physiological determinants of emotional state. *Psychological Review, 69,* 379-399.

SCHERER, K. R. (2000). Emotion. In Hewstone, M. & Stroebe, W. (Eds.), *Introduction to social psychology: A European perspective.* (3rd ed.), Oxford: Blackwell.

SCHORR, A. (2001). Appraisal: The Evolution of an Idea. In K. R. Scherer, A. Schor & T. Johnstone (Eds.), *Appraisal Processes in Emotion – Theory, Methods, Research* (pp. 20-34). New York: Oxford University Press.

SELIGMAN, M. E. P. (1970). On the generality of the laws of learning. *Psychological Review, 77,* 406-418.

SKINNER, B. F. (1938). *The Behaviour of Organisms.* New York: Appleton-Century-Crofts.

SKINNER, B. F. (1953). *Science and human behaviour.* New York: MacMillian.

SILVIA, P. J. (2005). What is Interesting? Exploring the Appraisal Structure of Interest. *Emotion, 5* (1), 89-102.

SOLOMON, R. C. (2002). Back to the Basics: On the Very Idea of "Basic Emotions". *Journal for the Theory of Social Behaviour, 32* (2), 115-144.

SOLOMON, R. C. & STONE, L. D. (2002). On "Positive" and "Negative" Emotions. *Journal for the Theory of Social Behaviour, 32* (4), 417-435.

STADDON, J. E. R. (1983). *Adaptive behaviour and learning.* Cambridge: Cambridge University Press.

TOLMAN, E. C. (1932). *Purposive Behaviour in Animals and Men.* New York: Appleton-Century-Crofts.

VON BERTANLANFFY, L. (1950). The Theory of Open Systems in Physic and Biology. *Science, 111,* 23-29.

WOODWORTH, R. S. (1958). *Dynamics of Behaviour.* New York: Holt-Rinehart-Winston.

WYCKOFF, L. B. (1952). The role of observing responses in discrimination learning. *Psychological Review, 59,* 431-442.

www.ingramcontent.com/pod-product-compliance
Lightning Source LLC
Chambersburg PA
CBHW021939170526
45157CB00005B/2349